Rodrigo Mendonça de Oliveira

Scorpionism in the municipality of Palmas - TO

Rodrigo Mendonça de Oliveira

Scorpionism in the municipality of Palmas - TO

Period from 2007 to 2013

"I dedicate this work to my mother for all the support she has given me at every moment of my life, and to the city of Palmas/TO, which has welcomed me very well and for which I owe many thanks. With this work, I proudly leave the entire population of Palmas my contribution to its growth."

ACKNOWLEDGEMENTS

Sincere thanks to God, first and foremost, for giving me the privilege of having passed the Master's programme and, above all, for having come this far.

I would like to thank my mother Sandra Mara Mendonça, who has never left me in difficulty and has always looked after me.

To the many friends I made in Palmas/TO, a land that welcomed me well and that, wherever I go, will make me proud to have lived there.

I would like to thank the Master's programme in Environmental Sciences - CIAMB and the CIAMB professors.

To my supervisor Prof.ª Drª . Carla Simone Seibert, who taught me a lot during the years from 2013 to 2015 when I was studying for my master's degree.

SUMMARY

SUMMARY

There are around 160 species of scorpions in Brazil, and those responsible for serious accidents belong to the genus Tityus. The species that inhabit the state of Tocantins are not considered to be of medical importance, but there are a relatively high number of accidents. The main aim of this study is therefore to analyse the conditions that lead to scorpion accidents in the city of Palmas-TO between 2007 and 2013. The topic was chosen because the city is young and was planned in an environment with great biodiversity, since the state is part of the Legal Amazon. The urbanisation process in Palmas took place in a disorderly fashion, and the population occupied and still occupies places where there is a very diverse fauna, including scorpions, which leads to accidents. In view of this, data was collected from the Notifiable Diseases Information System (SINAN), at the Palmas epidemiological secretariat, to identify the regions where scorpion accidents occurred between 2007 and 2013, where the epidemiological profile of scorpion accidents was also analysed. The distribution of accidents between the sexes was practically the same, so there was no distinction in terms of preference or ease of attack by scorpions. The frequency of accidents and their mapping were analysed by separating the notifications by area (urban and rural). During the analysis period, 653 cases of scorpion accidents were reported, 556 of which were in the urban area of Palmas. In order to identify the species of scorpions that occur in the municipality, active searches were carried out in houses, areas of collective activity of the population or native areas of the region, the species T. confluens was the most collected, thus being responsible for the majority of accidents in the municipality. There is a need to improve the notification process and the flow of information related to the disease. Greater mobilisation and targeting of health services in priority areas is suggested in order to improve the effectiveness of scorpionism surveillance and control strategies in Palmas-TO.

Keywords: Accidents, Scopes, Epidemiology

CHAPTER 1

INTRODUCTION

Scorpions originated more than 400 million years ago. Their remarkable evolutionary and adaptive capacity has enabled these animals to withstand all major cataclysms. In order to survive for millennia, scorpions have adapted to the most varied types of habitat, from deserts to tropical forests and from sea level to altitudes of up to 4,400 metres. However, most species prefer tropical and subtropical climates (LOURENÇO, 1984; CARDOSO, 2003).

All of today's scorpions are terrestrial and can be found in the most varied environments, in hiding places near human dwellings, buildings and under railway sleepers. They look for dark places to hide, and their nocturnal habit is recorded for most species (CARDOSO, 2003). They are most active during the warmer months of the year (particularly during the rainy season). However, due to climate change, in some regions these animals have been active all year round. They are carnivores, feeding mainly on insects and spiders, making them an efficient group of predators of a large number of small animals, sometimes harmful to humans (CANDIDO, 1999). Their predators include mice, coatis, monkeys, frogs, lizards, owls, seriemas, chickens, some spiders, ants, lacraias and scorpions themselves (BRASIL, 2006).

This animal is an efficient predator thanks to its strong pedipalps and agility, and is important for population control of various invertebrates and small vertebrates (PRENDINI, 2005). They therefore play an important role in the ecological balance as predators of other living beings, but can cause accidents when in contact with humans. Accidents are related to their feeding habits, form of reproduction, proliferation of species and behaviour combined with circumstances generated by humans (SALUN, 2010). In urban areas, preventive measures can be adopted to avoid their proliferation, through control, capture (active search) and environmental management (NEVES, 2010).

The species are endowed with venom and use this substance to capture and immobilise their prey, but only 25 species are potentially harmful to humans (CORRÊA, 2005), out of a group of more than 1,500 species, only five are found in Brazil (BRASIL, 2001). The mortality rate from scorpion accidents is approximately 0.5 per cent of cases, and these result from cardiorespiratory arrest in the elderly and children, who are the most victimised in accidents (ARAÚJO, 2003).

The extreme fear of this animal exists due to the population's lack of knowledge. The black scorpion (Bothriurus araguayae), for example, which is abundant in some cities in the state of São Paulo, has been causing people to panic, despite pain being the only clinical manifestation caused by its sting (SOARES, 2002).

1.1 Morphological and functional characteristics of scorpions

The scorpion's body, as shown in figure 1, is divided into: a carapace (prosoma), which contains a pair of chelicerae (used for crushing food); a pair of pedipalps (claws or hands); four pairs of legs; an abdomen (opisthosoma), formed by a trunk (mesosoma) where the ventral side contains the genital operculum and the sensory appendages in the form of combs, which allow it to pick up mechanical and chemical stimuli from the environment. There are also the spiracles, which are the external openings of the lungs; the tail (metasoma) which has an artery called a telson at the end, which ends in a stinger used to inoculate the venom. The telson contains a pair of venom-producing glands that flow into two holes situated on either side of the tip of the stinger (MATTHIESEN, 1972).

Arthropods have an external skeleton - the exoskeleton, a hard, chitinous structure that covers their body. Arachnids are arthropods without antennae, with four pairs of thoracic legs and a pair of palps. They breathe through phyllotracheae, leafy lungs, like pages in a book. Their body is divided into a cephalothorax and an abdomen (SELMA, 2005).

However, scorpions differ from other arachnids in that they have long palps, as well as the characteristic long and dangerous tail. The palps function as large, powerful pincers, which can be used to grip and dominate their prey. They are very sensitive to touch and air displacement due to the presence of very long, fine bristles. They can also have a greater number of eyes than other arachnids, with some species having up to six pairs, although this is not common (PARDAL, 2003).

The two venom glands, one on each side of the telson, which is the last segment of the post-abdomen (the so-called "tail"), produce complex mixtures of mucus and various proteins, including neurotoxic ones, which interact with the components of other animals' nerve cells and are responsible for the toxicity of scorpion venoms. This solution remains in the glands until they are compressed by muscles, when it is expelled through a channel that ends near the sting. The stimulus for this to happen comes from the need to paralyse prey or the realisation of a dangerous situation, in which case all the venom stored in the glands is released (BRASIL, 2006).

Scorpions are viviparous animals. The gestation period varies, but generally lasts three months for the Tityus genus. During labour, the female raises her body and makes a "basket" with her front legs, leaning on her hind legs. The newborns climb onto their mother's back through the basket and remain there for a few days, when they make their first skin change. After a few more days, they leave their mum's back and start living independently. The period between birth and dispersal of the young varies greatly. For Tityus bahiensis and Tityus serrulatu it takes around 14 days. Scorpions change their skin periodically, in a process called ecdysis; the old skin is the exuvia. They go through a limited number of molts until sexual maturity, when they stop growing (MANUAL DE CONTROLE DE ESCORPIÕES, 2009).

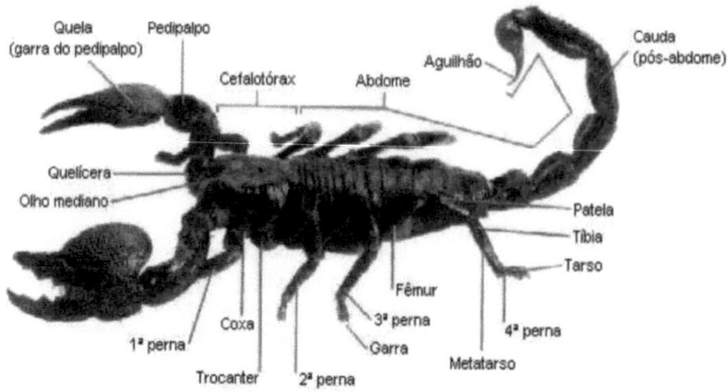

Figure 1 - Morphological and functional characteristics of scorpions
Source: www.geocities.ws

1.2 Scorpions and their Geographical Distribution in Brazil

In Brazil, there are around 160 species of scorpions. Those responsible for serious accidents belong to the Tityus genus, which is the richest in species, accounting for around 60 per cent of the Neotropical scorpion fauna, which are Tityus serrulatus, T.bahiensis, T.stigmurus and T.paraensis. The first three species, the yellow scorpions, are the most responsible for human accidents (SELMA, 2005). The most abundant species found in the state of Tocantins are: Tityusconfluens, Rhopalurus agamemnon, Tityus mattogrossensis, Tityus obscurus and Opisthacanthus cayaporum (CARDOSO, 2003).

The description of the species of medical importance in Brazil is as follows: Tityus serrulatus, also called the yellow scorpion, can reach up to 7cm in length, has a dark trunk, yellow legs, pedipalps and tail, which is serrated on the dorsal side, has light yellow legs and tail, and a dark trunk. The species is named after the presence of a serration on the 3rd and 4th ring of the tail. Its reproduction is parthenogenetic, in which each mother has approximately two births with around 20 offspring each year, reaching 160 offspring in her lifetime. It is considered the most poisonous scorpion in South America and is the cause of serious accidents, especially in the state of Minas Gerais. Tityus bahiensis is generally dark brown in colour, sometimes reddish-brown, with yellowish legs with dark spots. The femurs and tibiae of the pedipalps have dark spots. This scorpion causes the most frequent accidents

in the state of São Paulo. Tityus stigmurus is generally light yellow in colour with a black triangle on the head and a median longitudinal stripe and lateral spots on the trunk (MANUAL DE DIAGNOSTICO E TRATAMENTO DE ACIDENTES POR ANIMAIS PEÇONHENTOS, 1998).

Tityus cambridgei, on the other hand, is generally reddish-brown in colour, with light-coloured spots. The male has a longer tail than the female. The Tityus trivittatus is dark yellow in colour, with three almost black longitudinal stripes, and there may be slight variations in colour, reaching a size of around 7cm (MANUAL DE DIAGNOSTICO E TRATAMENTO DE ACIDENTES POR ANIMAIS PEÇONHENTOS, 1998).

1.3 Clinical Manifestations and Epidemiology of Scorpion Accidents

The neurotoxins produced by scorpions act mainly by stimulating peripheral nerve endings (nerves close to the skin), which causes a sensation of pain. In most cases, the symptoms are restricted to this point. In more serious cases, the victim's own body releases substances called chemical mediators, such as adrenaline, noradrenaline and acetylcholine, which have other serious consequences. Adrenaline and noradrenaline cause an increase in blood pressure, heart rhythm disturbances (cardiac arrhythmia), a reduction in the diameter of blood vessels (peripheral vessel constriction) and, eventually, heart failure, the accumulation of fluid in the lungs (pulmonary oedema) and, finally, shock, with a total disorder of the circulatory system and its inability to irrigate all the body's tissues. Acetylcholine causes an increase in tear, nasal, salivary, bronchial, sweat and gastric secretions, tremors, involuntary muscle contractions (muscle spasms), a decrease in pupil diameter (miosis) and a decrease in heart rate (TORRES, 2002).

According to the Ministry of Health (2001), cases of scorpionism can be classified into three levels according to their clinical severity: mild, when the symptoms are pain, local paresthesia (such as a sensation of cold, heat, tingling, among others) and there is no need to take an IV; moderate, when the symptoms are intense local pain associated with one or more other manifestations, such as nausea, vomiting, sweating, salivation, agitation, increased respiratory and cardiac rhythms, with the need to take an IV. In severe cases, there is profuse and incoercible vomiting (which cannot be controlled), intense sweating and salivation, prostration, convulsion, coma, reduced heart rate (bradycardia), heart failure, fluid accumulation in the lungs (acute pulmonary oedema) and shock.

Since accidents caused by venomous animals are a public health problem, their importance can be seen in the more than 100,000 accidents and almost 200 deaths recorded each year as a result of different types of poisoning (ARAÚJO, 2003). In Brazil, over time, records of these accidents have become more frequent, not only in hospital documentation, but also in newspaper articles and academic works (PARDAL, 2003). Of these, scorpionism has become increasingly common, accounting for 30 per cent of notifications in 2007, and surpassing cases of snakebite in absolute numbers (CANDIDO, 1999).

According to data from the Information System for Diseases and Notification (SINAN), 822 cases of accidents involving scorpions were reported in the state of Tocantins in 2010 and 2011. Palmas was the municipality with the highest number of records[1]. It is therefore necessary to analyse the conditions that are causing this rate in the capital, correlating accidents to environmental, urban, social, economic and cultural factors in the affected areas. To this end, it is necessary to identify and understand the distribution of scorpions prevalent in the different regions of the capital, in order to plan and scale the most appropriate strategies for action that will enable the biological control of these animals.

1.4 The Capital of Tocantins: Palmas

Palmas is the largest city in the state of Tocantins, founded on 20 May 1989, shortly after the creation of Tocantins by the 1988 constitution, with a population of 257,904 (IBGE, 2013).

The city of Palmas was conceived with the ideal of being a totally new, modern

---

[1] Data obtained from the Scientific Initiation report by Leobas, G. Accidents by venomous animals in the State of Tocantins: clinical-epidemiological aspects, 2013. Federal University of Tocantins.

city with an ecological and humanist vision. This would make it a hub for economic and social development for the entire state. A 90 kilometre by 90 kilometre quadrilateral was delimited between the municipalities of Porto Nacional and Taquaruçú, on the right bank of the Tocantins River, where the Palmas pilot plan was then implemented (SEDUH, 2002).

According to the Municipal Secretariat for Urban Development and Housing of Palmas - SEDUH (2002), Palmas was the last planned city built in the last millennium. It has its own characteristics, common to bold architecture, spaces previously designed by a modern urban layout and masterplan, aimed at the quality of life and comfort of its population.

The urban design of the city of Palmas, which was the subject of a previous urban project, took on a different form from what was thought and planned on paper. This shows the distance that exists between planning and public management of cities.

The original masterplan was based on an analysis of the environment and climate in the region, to ascertain the topography and soil conditions best suited to building the city. Thus, steep slopes, areas subject to flooding or erosion and places where soil treatment would require uneconomical solutions at the time of construction were eliminated. At the same time, the best possible use was made of breezes, landscapes and urban infrastructure (SEDUH, 2002).

> Since the establishment of the capital, with the promulgation of the State Constitution on 5 October 1989, a process of degradation and deforestation of Environmental Protection Areas and existing native vegetation has been observed. This degradation resulted from the urbanisation process, with the opening up of new blocks and the implementation of the road system. As a result, a large part of the natural areas were cut down, jeopardising the environmental quality of Palmas (BRASIL, 2006).

As a planned city, it was materialised as a new paradigm for urban hierarchisation in the interior of the Cerrado and the Amazon Portal. However, it has the same problems as other Brazilian cities in its segregationist, homogenising and hegemonic aspects (SANTOS, 2005). Its planners believed that through public-private partnerships Palmas would build itself. However, the masterplan was dismantled and the urban fabric went beyond the planned city in such a way as to develop contradictions between planning and management, due to the imposition of regional policies and speculation regulated by private capital (CARVALHÊDO, 2007).

From 1993 onwards, there was a disorderly occupation near the main sectors of the urban perimeter, generating an appearance of chaos in the planned city, and the formation of ghettos, with the construction of urban facilities in a palliative manner. With the occupation of immigrants outside the master plan, the city became dehumanised, generating a feeling of not belonging to the city (LIRA, 2005).

The process of urban occupation in Palmas has had a strong impact on the quality of life of its residents and has aroused interest in studies that can carry out investigations based on indicators that aim to measure important aspects of the quality of urban life in this city (SEDUH, 2002).

The highest densities in the South Palmas region (Taquaralto and Aureny's Gardens) and in the North Region blocks coincide with the areas initially occupied through urban plot invasions (some of which were later urbanised and regularised), where the poorest population migrated, with this trend of occupation continuing in the other blocks of this region of the city and the emergence of peripheral neighbourhoods (SEDUH, 2002).

Some of the blocks in the Palmas Centro region (many of which are characterised by permanent housing with good construction standards) are also unpaved, causing problems for the population, such as the difficulty for rubbish collection vehicles to circulate, especially during the rainy season, and the occurrence of respiratory diseases due to the intensity of the dust during the dry season. The difficulty of access for rubbish collection vehicles influences the habits of residents, who, not having access to this public service, tend to throw waste on vacant lots or even on public roads, compromising the environmental quality of the city (KRAN; FERREIRA, 2006).

The rubbish collection system in the urban area of Palmas is run by the City Council, which serves around 90 per cent of the city's households through periodic collection on three alternate days a week (SEDUH, 2013). The service has been considerably extended throughout the urban area, with no significant differences between the blocks in the centre and the blocks and neighbourhoods in the outlying sectors.

1.5 Characterisations of the Vegetation of Palmas

To understand the species of scorpions that exist in Palmas a little better, it is necessary to understand what kind of habitat they prefer, and Palmas favours everything that these species need for their development according to the Scorpion Control Manual (2009).

The municipality of Palmas is located in climate regionalisation C2wA'a' - characterised as a humid sub-humid climate with moderate water deficiency in winter, average annual potential evapotranspiration of 1,500 mm, distributed in summer at around 420 mm over the three consecutive months with the highest temperature (SEPLAN, 2008).

Palmas has different phytophysiognomies with countryside characteristics, according to the IBGE (2013). It is worth noting that countryside areas are defined as different categories of vegetation that are physiognomically quite different from forests, i.e. those characterised by a predominantly shrub layer, sparsely distributed over a grassy-woody carpet.

According to the Technical Manual of Brazilian Vegetation (2012), this category includes the characteristics of the Grassy-Ligneous Savannah (Campo-Limpo-de-Cerrado), which prevail in this physiognomy. When natural, the lawns are interspersed with stunted woody plants, which occupy extensive areas dominated by hemicryptophytes and which, little by little, when managed through fire or grazing, are being replaced by geophytes that are distinguished by having underground stems and are therefore more resistant to trampling by cattle and fire.

In the study area, this phytophysiognomy characterised by the presence of shrubs and subshrubs is insignificant, which allows us to conclude that this class is equivalent to low vegetation without gallery forest, subject to annual fire.

The vegetation of the municipality of Palmas is located in the central area of the Brazilian Savannah (Cerrado), conceptualised as a xeromorphic vegetation, which occurs under different types of climate and is characterised by having two distinct components: the shrub and the herbaceous-subshrub (IBGE, 2013). This favours the proliferation and development of the species that inhabit the state of Tocantins.

CHAPTER 2

GENERAL OBJECTIVE

- Analysis of scorpionism in the municipality of Palmas/TO, between 2007 and 2013.

2.1 SPECIFIC OBJECTIVES

- Analysis of the epidemiological profile of scorpion accidents in the municipality of Palmas/TO;

- Mapping the locations of the highest incidence of accidents in the district of Palmas/TO;

- Identification of the species that cause accidents in the municipality of Palmas/TO;

CHAPTER 3

METHODOLOGY

The epidemiological study of scorpion accidents in Palmas was based on data from the Notifiable Diseases Information System (SINAN), obtained from the Health Department, in the Epidemiology sector of the municipality of Palmas. As notifications of scorpion accidents in Palmas were made more accurately from 2007 onwards, due to the modernisation of the IT system, the research covered this period up to 2013.

In order to analyse the epidemiological profile of scorpion accidents, information was taken from all the notifications made for the municipality, regardless of the location of the accident.

The information analysed was related to 1) the injured individual (age, sex and place of the accident); 2) the condition triggered by the accident, such as the site of the bite, time taken to receive care, classification of the accident (mild, moderate and severe), whether there was treatment with serotherapy and the evolution of the cases. The results were tabulated, separated by year of occurrence and analysed descriptively .

The notifications were grouped by area of occurrence, considering urban and rural areas. The notifications for the urban area were separated into the Headquarters District (Central and Southern Regions) and the Other Districts (Taquaruçu and Buritirana Districts). For the Central region, we considered the notifications registered for the North sector - blocks above Avenida JK; and the South sector - blocks below Avenida JK (Figure 2).

The notifications from the South Zone of Palmas were those whose accidents occurred in the neighbourhoods of Taquaralto, Aureni I, Aureni II, Aureni III, Aureni IV, Santa Fé, Morada do Sol, Bela Vista, Santa Bárbara, Setor Sul, Vale do Sol, Maria Rosa, Sol Nascente, Lago Sul, Setor Universitário, Setor Aeroporto, Setor Industrial, Santo Amaro, Água Fria, Marly Camargo, Taquari, Santa Helena, União Sul and Irmã Dulce (Figure 2).

In order to assess the evolution of scorpion accidents over the years under study, we analysed the notifications that occurred only in the municipality's Main District, based on the years 2007, 2010 and 2013. Notifications from the districts of Buritirana and Taquaruçu were not included in this analysis.

To assess the evolution of land use and land cover over the incidence of accidents, thematic maps were drawn up using Landsat satellite images in bands 3, 4 and 5 from IMPE, the National Institute for Space Research, based on the years 2007, 2010 and 2013. The thematic classes used were Field; Cerrado in the Restricted Sense; Cerradão; Gallery Forest/Ciliary Forest; Urbanised Area; Beach; and Continental Bodies of Water; and Farming.

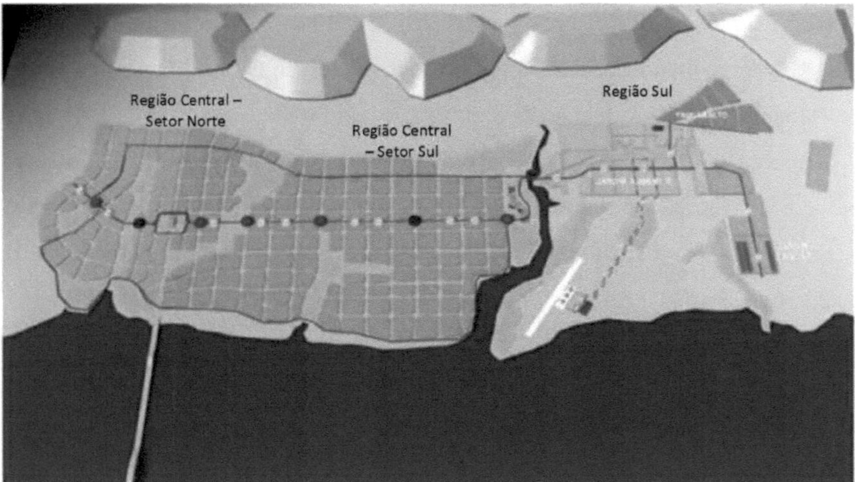

Figure 2-Distribution diagram of the blocks and neighbourhoods of the Palmas - TO Municipal District, according to the Master Plan
Source: Nortista (2014)
2

Legend: Central Region (North Sector: orange blocks; South Sector: blue blocks), South Region (yellow and brown blocks).

In order to identify the species of scorpions that occur in the municipality, active searches were carried out in houses, areas of collective activity of the population or native areas of the region, from August 2013 to August 2014, by the team from the municipality's Zoonosis Control Centre (CCZ). The species collected were stored in

---

[2] Available at: <http://www.skyscrapercity.com/showthread.php?t=1707463&page=2>.

vials and the location and date of collection were recorded. Species identification was carried out in the Zoonosis Control Centre laboratory, based on the catalogue in the Ministry of Health's Scorpion Manual (2009).

With this data, a distribution map of the species found in the municipality was drawn up. Images of the scorpion species collected were recorded, as well as some urban areas considered to be at risk in the municipality for the proliferation of these animals.

The maps were drawn up using Esri/ArcGis 10.2 software. The information plans (IPs) were edited from the Digital Cartographic Base for the blocks of Palmas, made available in DWG format on the Palmas City Hall website via the SIG-PALMAS link. In this procedure it was necessary to convert the data from DWG format to shapefile/ESRI format, and at the end of this stage a Projection System was added: Universal Transverse Mercator UTM with Brazilian Geodetic Datum SIRGAS 2000 / Zone: 22 South, so that it could be edited in the Geographic Information System (GIS).

CHAPTER 4

RESULTS

In order to carry out the study, 653 scorpion accidents were analysed and reported on the Notifiable Diseases Information System (SINAN) by hospitals and health centres in the municipality of Palmas. There was an increase in the number of accidents reported over the seven years analysed, with 22% of these reported in 2013 alone, 113 more than in 2007 (Table 1).

Scorpion accidents occurred in very similar proportions for both sexes. The highest incidence was recorded among people aged 20 to 49 (358 accidents), but accidents involving children (105 cases) and the elderly (2 cases) were also reported during the period. The majority of accidents (550) occurred outside the workplace, either at the injured person's home or during leisure time (Table 1).

The clinical picture of patients injured by scorpions is shown in Table 2. The hands (176 cases) and fingers (124 cases) were the most affected parts of the body. For 422 accidents, the time taken to receive medical attention was not recorded; for a further 149 cases, the patient received medical attention within 1 hour of the accident. Of the 653 accidents, only 6 were classified as serious, the rest as moderate (262 cases) or mild (385 cases). Of those injured, 301 were treated with serotherapy and the majority progressed to a cure, without sequelae (649 cases).

Table 3 shows scorpion accidents, separated by area of occurrence and region of the municipality of Palmas. Of the 653 accidents reported between 2007 and 2013, 556 (85% of cases) occurred in the urban area and of these, 230 were in the southern region of Palmas.

Table 1 - Profile of patients injured by scorpions in the municipality of Palmas (TO) between 2007 and 2013

-	2007	2008	2009	2010	2011	2012	2013	Total
Sex								
Male	23	33	65	45	43	58	65	*332*
Female	15	26	55	43	52	50	80	*321*
Age group								
<1-9	3	15	21	9	10	24	23	*105*

								Total
10-19	8	15	20	13	17	20	26	*109*
20-34	16	17	37	42	33	38	55	*238*
35^9	5	7	34	14	21	15	24	*120*
50-64	4	5	**6**	8	11	9	13	*56*
65-79	2	0	2	2	2	2	3	*13*
80 e +	0	0	0	0	**1**	0	1	*2*
Accident in				***work?***				
Yes	11	6	23	12	17	17	17	103
No	27	53	97	76	78	91	128	550
Total	**38**	**59**	**120**	**88**	**95**	**108**	**145**	**653**

Source: SINAN (2014)

The map of the evolution of scorpion accidents in the municipality of Palmas was constructed using notifications for the urban area, excluding district records, based on the years 2007, 2010 and 2013 (Figure 3). This showed an increase in the number of reports of scorpion accidents and the locations where these accidents occur. When analysing the accidents notified for the Central region in the North Sector, in 2010 there was an increase in the number of accidents, mainly in ARNO 33, 43, 44, 73, 72 and 71. In 2013 there was the highest number of notifications for these accidents, distributed more widely throughout the sector (Annex 1 and 2).

For the Central Region in the South Sector, there were 8 accidents in 2007, 26 in 2008 and 31 in 2009. In 2010 and 2011 there was a reduction in accidents, but these increased again in the following years, 2012 and 2013.

As for the southern zone, few cases were reported in 2007, with accidents in Taquaralto and three cases reported in the Taquari neighbourhood. Since 2011, the Taquari neighbourhood has reported the highest number of accidents (Annex 1 and 2).

Table 2 - Clinical profile of scorpion accidents in the municipality of Palmas -TO, from 2007 to 2013

	Local sting							
	2007	**2008**	**2009**	**2010**	**2011**	**2012**	**2013**	**Total**
Arm	1	3	3	4	3	5	6	*25*
Anie-Braço	1	1	0	2	3	3	3	*13*
Hand	7	13	34	26	29	28	39	*176*
Finger	14	14	23	15	9	19	30	*124*
Trunk	1	4	7	4	4	7	5	*32*
Thigh	1	1	5	3	6	7	6	*29*
Leg	1	1	10	8	11	7	8	*46*
Foot	9	16	33	19	24	21	37	*159*
Toe	3	3	2	6	4	7	10	*35*

Service time							
Ign/White	18	37	72	63	57	77	98	*422*
0-1 hours	15	16	27	13	26	20	32	149
1-3 hours	4	1	14	10	8	8	7	52
3-6 hours	0	1	3	2	3	1	3	13
6-12 hours	1	3	0	0	1	1	**0**	6
12 and +	0	1	4	0	0	1	5	**11**
hours							
Classification of accidents							
Lightweight	22	41	81	34	57	64	86	*385*
Moderate	15	16	37	54	38	43	59	*262*
Grave	1	2	2	0	0	1	**0**	6
Applied serotherapy							
Ign/White	0	0	0	0	0	0	1	*1*
Yes	20	21	60	57	42	44	57	*301*
No	18	38	60	31	53	64	87	*351*
Evolution of cases							
Healing	*0*	*0*	*0*	*0*	*0*	*0*	*4*	*4*
Cure/sequel	38	59	120	88	95	108	141	649
Total	**38**	**59**	**120**	88	95	**108**	**145**	**653**

Source: SINAN (2014)

Table 3 - Distribution of scorpion accidents, by zone and region of the municipality of Palmas-TO, between 2007 and 2013

	2007	2008	2009	2010	2011	2012	2013	Total
Distribution of accidents by zone								
Urban Area	33	53	99	75	75	91	130	556
Rural Area	5	6	21	10	19	15	21	97
Total	**38**	**59**	**120**	**85**	**94**	**106**	**151**	**653**
Distribution of accidents in the urban area								
Head Office								
Central Region-North Sector	12	12	20	14	15	17	20	110
Central Region-South Sector	9	27	32	22	25	23	43	181
Southern Region	11	13	42	31	30	48	55	230
Other Districts	1	1	5	8	5	3	12	35
Total	**33**	**53**	**99**	**75**	**75**	**91**	**130**	**556**

Source: SINAN (2014)

Figure 4 shows the map of the evolution of land use and land cover in the municipality of Palmas-TO for the years 2007, 2010 and 2013. In 2007, there were agricultural areas in the northern and southern sectors of the central region of Palmas, and significant areas of restricted savannah to the west in the southern sector of the central region and in the southern region of the municipality. Farming was replaced by restricted savannah or by urban areas in the following years, but there is a visible increase in urbanisation, especially in Jardim Taquari.

Figure 5 shows risk areas for the proliferation of scorpions in Palmas, such as rubble, poorly conditioned rubbish or rubbish accumulated in unoccupied areas, vegetation close to homes, among others.

With the active searches to collect scorpions, five species were found in the municipality of Palmas: Tityus confluens, Rhopalurus agamemnon, Tityus mattogrossensis, Tityus obscurus and Opisthacanthus cayaporum. Between 2013 and 2014, 55 scorpion specimens were collected, with Tityus confluens (32 specimens) and Rhopalurus agamemnon (15 specimens) predominating. One specimen of a probable Tityus obscurus, four specimens of Tityus mattogrossensis and one specimen of Opisthacanthus cayaporum were found, which is still being recognised by the CZZ. For two other specimens collected, it was not possible to identify the species. The image chart with the distribution map of these species is shown in figure 6 and their image in figure 7.

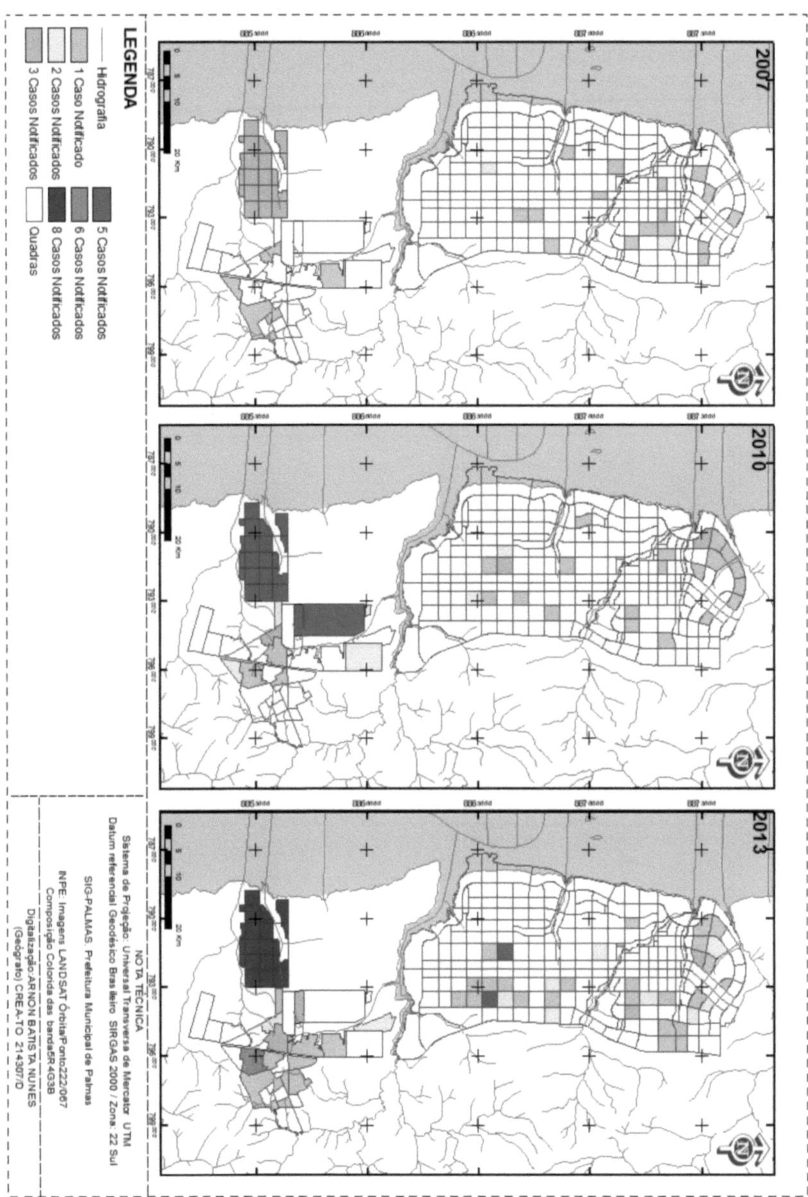

Figure 1 - Evolution of scorpion accidents in the district of Palmas-TO for the years 2007, 2010 and 2013
Source: Palmas (undated); SINAN (2014).

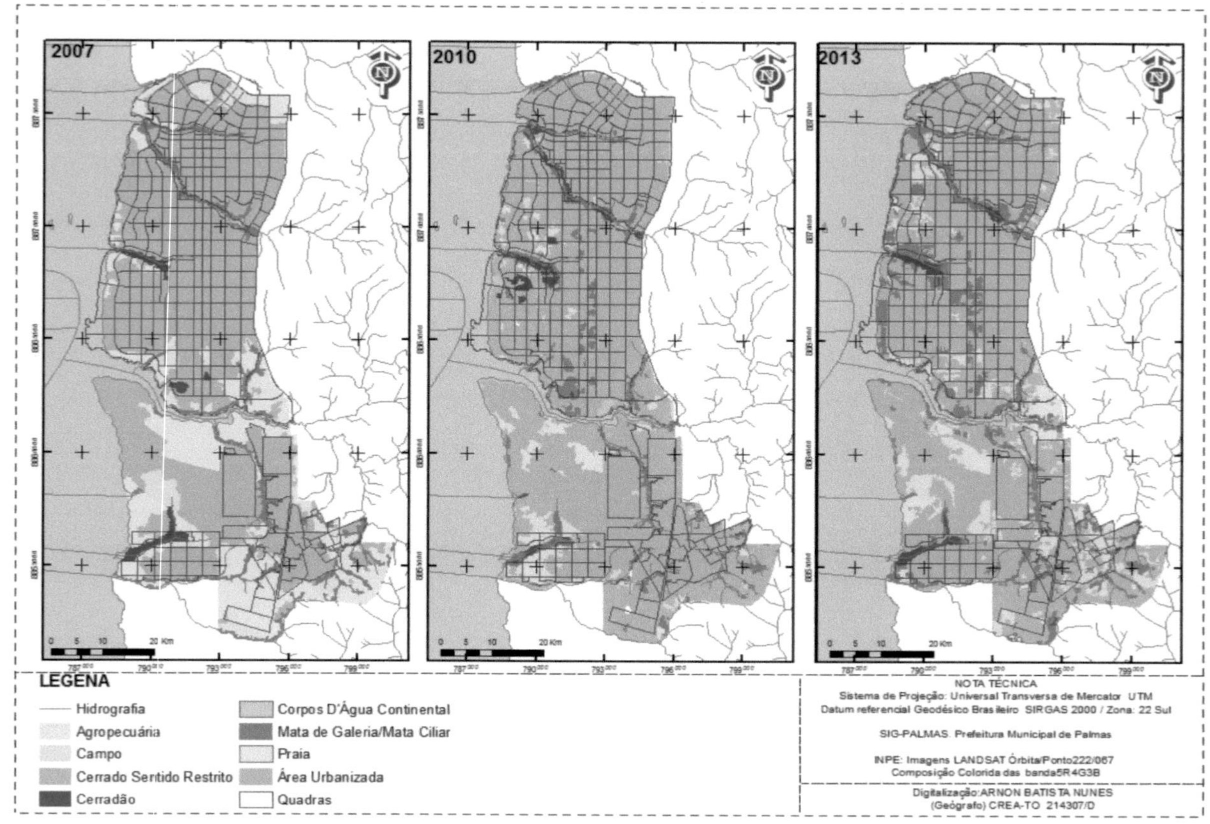

Figure 2-Evolution of land use and land cover in the district of Palmas-TO, for the years 2007, 2010 and 2013.

Source: Palmas (undated)

Figure 3- Areas of risk for scorpion proliferation in the municipality of Palmas - TO **A**. Quadra 108, North Sector; **B**. Quadra 104, South Sector; **C**. Quadra 107, South Sector; **D**. Aureni III neighbourhood, South Zone.

Source: Oliveira, RM (2015)

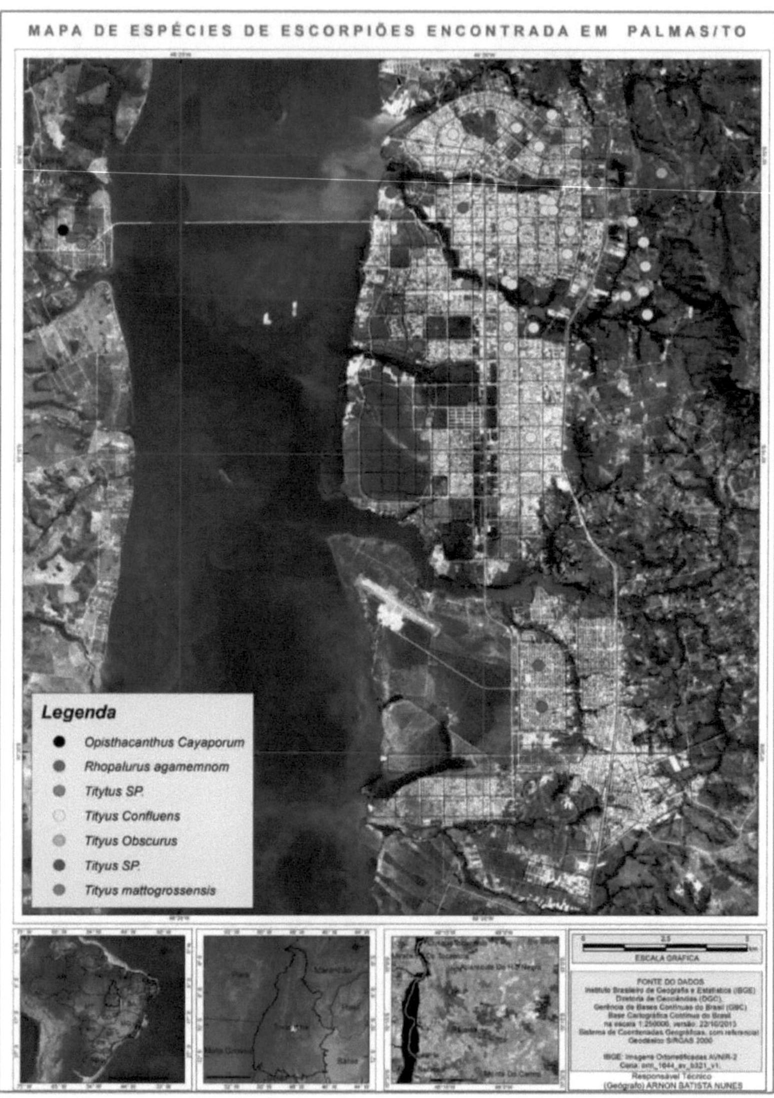

Figure 6 - Image map of the distribution of scorpions in the municipality of Palmas-TO, specimens collected in 2013 and 2014
Source: Zoonosis Control Centre (CCZ)

Figure 7-Species of scorpions collected in the municipality of Palmas-TO between 2013 and 2014 **A. Tityus confluens; B. Opisthacanthus cayaporum; C. Tityus mattogrossensis; D. Tityus obscurus; E. Rhopalurus agamemnon**

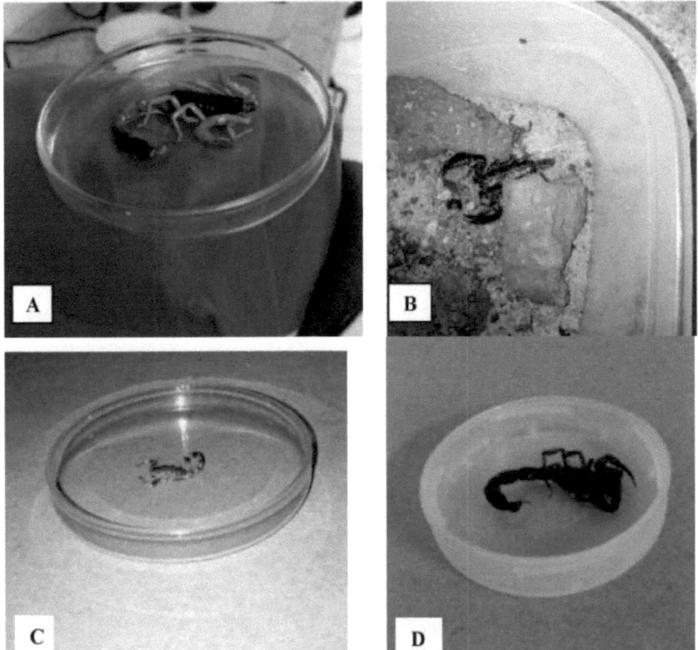

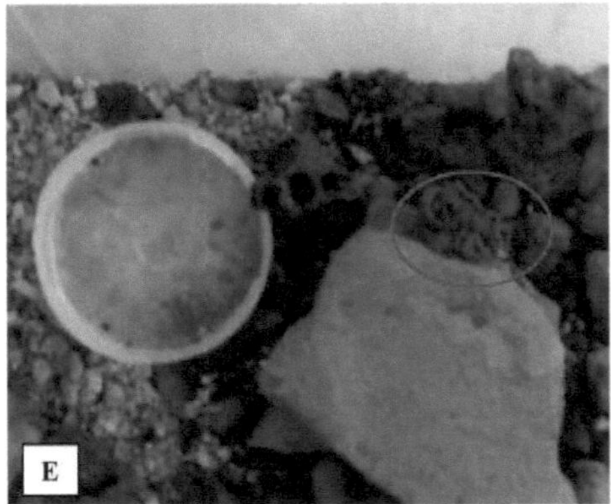

Source: Oliveira, RM (2015)

CHAPTER 5

DISCUSSION

In order to uncover the scorpion accidents that occurred between 2007 and 2013 in the municipality of Palmas-TO, this paper presents the profile of the people injured, the clinical condition triggered by the accident, the distribution of these accidents in the municipality, land occupation in the urban space and the species of scorpions collected between 2013 and 2014.

Scorpion accidents occurred in very similar proportions for both sexes, in a more active age group of the population and with a significant proportion of accidents occurring outside the workplace. Many of the accidents involving scorpions occur inside homes, as this is a place with a history of abundant cockroach populations and, consequently, one that favours the proliferation of these animals.

Scorpions can live anywhere, as long as they have something to eat. They can live in rural or urban areas, in homes or workplaces (LOURENÇO, 1980). These animals look for food at night and can enter environments through wiring and sewage pipes, as well as gaps in walls, doors and windows (PRENDINI, 2005). They can hide from daylight in dark places such as inside shoes, cupboards, drawers, cloths and towels, in laundry areas and bathrooms. As a result, the group most at risk of these accidents are children and housewives, who spend the longest period of time in or around the home.

Of the 653 accidents reported in these seven years of study, 16% (105 cases) involved children up to 9 years old, but 54% of accidents (358 cases) were reported for people aged between 20 and 49. This is because scorpion accidents can also happen to people who work in the construction industry, woodworkers, hauliers and fruit and vegetable distributors, because they handle objects and food where scorpions can be found (LOURENÇO, 2002).

The most notorious attribute of a scorpion is its sting/venom, although it is true that scorpions are among the most poisonous animals living on land. All scorpion species are poisonous to insects, which are their potential food. However, of the approximately 1,500 known species of scorpions, only a small number are dangerous to humans (CANDIDO, 1999). The venom of most scorpions in humans produces a reaction similar to that of a bee sting, which is very painful but in most cases not life-threatening (POLIS, 1990).

Although they don't attack humans intentionally, scorpion accidents occur when a person touches their hand, foot or other part of their body to the animal. Because they are very small, they can hide anywhere and remain motionless for hours and days, so all it takes is for someone who is distracted to touch the animal.

In this study, the number of accidents in which the sting occurred on the hand and finger shows an occasional accident and a lack of attention on the part of the victim in observing what they are handling. Cupo et al. (2003) developed a work on scorpionism, referring to the epidemiological issue, which cites a survey carried out in three cities. Although the species analysed in the work cited do not match those found in the municipality of Palmas, there are similarities with the data in relation to the site of the sting (73.7% of accidents occurred on the hands and feet), corroborating the perspective that there is inattention on the part of the population.

The large number of accidents without a record of the time taken to treat them (64 per cent, 422 cases), the low severity of the clinical condition of the victims and their progression to cure without sequelae, is in line with the species most commonly collected in the municipality, Tityus confluens and Rhopalurus agamemnom. These species trigger a milder clinical picture in accidents involving humans (LOURENÇO, 2002).

A scorpion accident is usually followed by pain (moderate or intense) or tingling at the site of the sting. These symptoms can be treated with analgesics or anaesthetics, but you should watch out for the appearance of other symptoms for at least 6 to 12 hours, especially in children under 7 and the elderly. Serotherapy for the treatment of these patients is carried out when the clinical manifestations are more severe, and in this study, 46% of the victims (301 cases) were treated with anti-scorpion serum, for 41% of the patients whose clinical condition was considered moderate or severe.

There was a growing increase in reports of scorpion accidents in the municipality of Palmas from 2007 to 2013. These accidents are the most reported in the municipality and surpass those caused by snakes, caterpillars and bees (SILVA, 2015).

Palmas is an ever-expanding capital city, with rapid growth and a number of construction sites scattered throughout the city, in both commercial and residential areas. As a result, there is a large accumulation of building materials and rubble, providing a suitable habitat for scorpions. On the other hand, the municipality's expansion has also been accompanied by socio-spatial segregation, which has led to an increase in problems such as the disorderly clustering of houses, basic sanitation, rubbish collection and the proximity of homes to vegetation (KRAN; FERREIRA, 2006).

It is possible to assess the urban expansion of the municipality of Palmas by analysing the land cover and land use map (Figure 4). It shows the transformation of agricultural and/or green areas into urbanised areas in all regions of the municipality, especially on the outskirts of the Main District. The outlying regions are also the ones closest to the forests.

The images taken in the different regions of Palmas highlighted unpaved streets, homes close to vegetation, building rubble and rubbish in inappropriate places, all situations that are favourable to the proliferation of scorpions (Figure 5). In this way, the increase in scorpion accidents over the years seems to be linked to the urban growth of Palmas, in all regions of the municipality, especially the North Sector of the Central Region and the South Region.

The southern region of Palmas had the highest number of accidents (41%, 230 cases). This part of the municipality was not included in the master plan, there was no planned organisation in this area and it is also where the highest percentage of people on low incomes are located (KRAN; FERREIRA, 2006; ARAÚJO et al., 2003). The neighbourhood most affected by accidents in this region was Taquari (40 cases). The high accident rate can be explained by the fact that it is a neighbourhood structured in a former rural area and with poor infrastructure conditions, which tend to lead to the accumulation of rubbish and debris. Other neighbourhoods in the Southern Region also had high rates of scorpion accidents, such as Aurenis (I, II, III and IV) and Taquaralto.

Thus, although Palmas is a planned and young city, it has several peripheral areas with limited accessibility and infrastructure, where the land is less valued and occupied basically by low-income populations, who live in evident conditions of economic, social and environmental precariousness, and consequently have a compromised quality of life (KRAN; FERREIRA, 2006).

With the scorpion collection carried out in the municipality, five species were identified: Tityus confluens, Tityus mattogrossensis, Rhopalurus Agamemnon, Opisthacanthus cayaporum and Tityus obscurus. All of these species have been reported in the state of Tocantins and have been described for areas of cerrado and the transition between savannas and the Amazon rainforest (LOURENÇO, 2002; BRANDÃO, 2005; BRASIL, 2009), characteristics that are also present in the municipality of Palmas. It should be noted that only one specimen of Tityus obscurus was collected by a resident, who handed it over to the CCZ. This species had not yet been identified in Palmas and so it is believed that it is not responsible for accidents in the municipality.

It is important to emphasise that none of the species found in the municipality cause serious accidents to humans, and are therefore not reported in the literature as species of medical interest, as is the case with Tityus serrulatos and Tityus stigmurus.

Palmas is a city under construction whose vegetation is being modified by human occupation. The removal of the vegetation causes a break in the scorpions' food chain, also putting an end to their places of shelter. With the scarcity of food in the native area, these animals start looking for food and shelter in homes, vacant lots and construction areas. Places where there is an accumulation of organic matter, rubble, rubbish, deposits and warehouses, conditions that favour the proliferation of cockroaches (Periplaneta americana, and other species) due to the availability of food and humidity (LOURENÇO, 2002). As scorpions have cockroaches as components of their food chain, they move to places where they are abundant, getting closer to the population.

Therefore, the increase in reports of scorpion accidents over the years under study is related to the changes that have taken place as a result of the urban expansion of the municipality, which involves alterations to the environment, the socio-economic and cultural conditions of the population and the urban planning of the city. It is therefore necessary for public management to develop actions to minimise the conditions in which these animals proliferate.

CHAPTER 6

FINAL CONSIDERATIONS

Scorpionism in Palmas is a reflection of disorganised urban expansion, which should be planned, following the guidelines of the Master Plan for the entire urban space. Today, the incidence of scorpion accidents is growing and requires action to combat their proliferation. Municipal managers need to direct their efforts towards guiding the population to control scorpions, informing them of the importance of taking care to clean the environment, properly storing rubbish, and also the need to close off spaces that allow the animal to enter the home. With this, basic actions can reduce the number of people injured by scorpions.

The definition of priority areas should be the basis for mobilising health services to create forward-looking plans to improve the effectiveness of surveillance and control strategies for the disease in the municipality. These actions, in turn, should be carried out mainly in the form of active searches immediately after notification of the accident and spontaneous demand from the population for effective awareness-raising work, as well as visits to risk areas and neighbouring areas. When planning these actions, the differences within each region should also be taken into account, in terms of the level of information and socio-economic status of residents, environmental conditions and the type of use and occupation of space.

It is also important to emphasise the need to report cases correctly, identifying the agent that caused the accident and filling in the SINAN form completely, which will help epidemiological studies to show the reality of the municipality.

CHAPTER 7

REFERENCES

ANDRADE, F.C.D. (2000) The evolution of social mobility in five Brazilian metropolitan regions, 1988 and 1996. In: ENCONTRO NACIONAL DE ESTUDO POPULACIONAIS, 12, 2000, Caxambu, *Anais.* Caxambu: ABEP.

ARAÚJO, F. A., SANTALÚCIA, M. & CABRAL, R. F. Epidemiology of Accidents by Venomous Animals. In: *Animais Peçonhentos no Brasil: Biologia, Clínica e Terapêutica dos Acidentes* (J. L. Cardoso et al.), 2003.

BORELLI, A. 1899. Travels of Dott. A. Borellinella Republica Argentina en el Paraguay. XXIII. Scorpioni. *BollettinodeiMusei di zoologia e d'anatomia compara tadella real Università di Torino*, v.14, n.336, p. 1-6, 1899.

BRANDÃO R.A., Motta P.C. *Circumstantial evidences formimicry of scorpions by the neotropical gecko Coleo dactylus brachystoma (Squamata, Gekkonidae) in the Cerrados of central Brazil.* Phyllomedusa; v. 4, p.139-145, 2005.

BRAZIL. Law No. 5.197, of 3 January 1967. Provides for the protection of fauna and other measures. *Official Gazette of the Federative Republic of Brazil*, Brasília, DF, 7 January 1967.

____________ . *Manual for the Diagnosis and Treatment of Accidents by Venomous Animals.* Ministry of Health and National Health Foundation, 1998.

____________ . *Manual of Diagnosis and Treatment of Accidents by Venomous Animals.* Brasilia: Ministry of Health. National Health Foundation, 2001.

BRASIL. *Spatial approaches in public health.* SANTOS, S. M. ; BARCELLOS, C. (org.) Ministério da Saúde, Fundação Oswaldo Cruz;- Brasília: Ministério da Saúde, 2006 (Série B. Textos Básicos de Saúde) (Series for Training and Updating in Geoprocessing in Health; 1).

. Normative Instruction No. 141 of 19 December 2006. Regulates the control and environmental management of harmful synanthropic fauna. *Official Gazette of the Federative Republic of Brazil*, Brasília/DF, 19 December 2006.

____________ . *Manual de controle do Escorpião.*Ministério Da Saúde, Secretaria de Vigilância em Saúde, Departamento de Vigilância Epidemiológica, Brasilia/DF, 2009. (Series B. Basic Health Texts. Print run: 1ª edition - 2009 - 20,000 copies).

CANDIDO, D. M. Scorpions. *In:* BRANDÃO, C. R. F.; CANCELLO, E. M. (Ed.).

Biodiversity of the State of São Paulo: Terrestrial Invertebrates. São Paulo: FAPESP, v.5, p. 23-34. 1999.

CARDOSO, J. L. C. José de Anchieta and the Letters. In: CARDOSO, J. L.; et al. (org).*Animais Peçonhentos no Brasil: Biologia, Clínica e Terapêutica dos Acidentes*. São Paulo: Sarvier. 2003.

CARVALHÊDO, W. S. Segregação Urbana: uma análise sócio-espacial da capital Palmas/TO. Porto Nacional: Article (Undergraduate Degree in Geography) - Federal University of Tocantins, 2007.

CORRÊA, R. L. O Espaço Urbano. 4 ed. São Paulo: Ática, 2000.

BRAZILIAN INSTITUTE OF GEOGRAPHY AND STATISTICS. Technical manual of Brazilian vegetation. 2. ed. rev. and ampl. Rio de Janeiro: IBGE, 2012. 271 p. (Technical manuals in geosciences , n. 1). Available at : <http://geoftp.ibge.gov.br/documentos/recursos_naturais/manuais_tecnicos/manual_tecnico_vegetacao_brasileira.pdf>. Accessed on 12 December 2013.

KRAN, F.; FERREIRA, F. P. M. *Qualidade de vida na cidade de palmas - to: uma análise através de indicadores habitacionais e ambientais urbanos*. Ambiente & Sociedade, v. IX, n 2, jul./dez. 2006

LIRA, Elizeu Ribeiro. A Produção do Espaço de Palmas: "comprometer para desenvolver" Produção Acadêmica, Porto Nacional, v. 2, n. 2. p. 157-171, jan. 2005.

LIRADASILVA, R. M.; AMORIM, A. M.; BRAZIL, T. K. Poisoning by Tityus stigmurus (Scorpiones, Buthidae) in the State of Bahia, Brazil. Revista da Sociedade Brasileira de Medicina Tropical, [S.l.], v. 33, n. 3, p. 239-245, 2000.

LOURENÇO, W. R. 1980. Contribution to the comparative study of Scorpions appartenant to the "complex" *Tityus trivittatus* Kraepelin, 1898 (Buthidae).*Bulletin Du Muséum national d'Histoire naturelle,* Paris, 4e ser., 2(A3): 793843.

LOURENÇO, W. R. Critical review of the Tityus species of the State of Pará (Scorpiones, Buthidae). Boi. Mus. Para. Emilio Goeldi, ser. Zool., v.l, n.l, p.5-18, abro 1984.

LOURENÇO, W.R. 1988. Biographical, ecological and evolutionary considerations on neotropical species of Opisthacanthus Peters, 1861 (Scorpiones, Ischnuridae). Studies on Neotropical Fauna and Environment, 23(1): 21-53.

LOURENÇO, W.R., CLOUDSLEY-THOMPSON, J.L., CUELLAR,O. The evolution of scorpionism in Brazil in recent years.J Venom Anim Toxins, 1996.

LOURENÇO, W.R. 2002.*Scorpions of Brazil.Les* Editions de l'IF, Paris, 307pp.

MATTHIESEN, F. A. *The Scorpion*. São Paulo: Edart, 1976

MAURY, E. A. 1974. Chaquena scorpion fauna. II. *Tityus confluens* Borelli 1899 (Buthidae). *Physis*(C), a. 33, v.86, p. 85-92.

MELLO LEITÃO, C. South American scorpions. Archives of the National Museum, Rio de Janeiro, a.40, p. 1-468.

MORIN, E. Ciência com Consciência Edition revised and modified by the Author Translation Maria D. Alexandre and Maria Alice Sampaio Dória 82 EDIÇÃO, EDITORA BERTRAND BRASIL LTDA.

MS (Ministry of Health) / FUNASA (National Health Foundation). Manual for the Diagnosis and Treatment of Accidents by Venomous Animals. Brasília: MS/FUNASA. 2001.

MS (Ministry of Health), Department of Epidemiological Surveillance. Scorpion Control Manual. Ministry of Health. Brasília; 2009. p. 72. 5.

NEVES. D. A. P. Preliminary Results of the 2010 Demographic Census Universe.

PALMAS. Review of the Palmas Master Plan IPUD.Secretaria Municipal de Desenvolvimento Urbano e Habitação SEDUH ,Palmas, 2002.

PALMAS. Participatory Masterplan for Territorial Development - Palmas - TO. Secretaria Municipal de Desenvolvimento Urbano e Habitação/ Universidade Federal Do Tocantins - UFT/ Assoc. Para a recuperação e a conservação do ambiente - arca,Palmas 2006. 82 p.: il. - (Specialisation Course in Urban and Environmental Planning) (Environmental Report of Technical and Community Reading of the Municipality of Palmas TO).

______________. Participatory Masterplan for Territorial   Development - Palmas - TO. Secretaria Municipal de Desenvolvimento Urbano e Habitação/ Universidade Federal Do Tocantins - UFT/ Assoc. Para a recuperação e a conservação do ambiente - arca, Palmas 2006. 29 p.: il - (Specialisation Course in Urban and Environmental Planning) (Analysis of Urban Occupation in Palmas).

PARDAL, P. P. O. etal.Aspectos Epidemiológicos e clínicos do escorpionismo na região de Santarém,Estado do Pará, Brasil.In:R. Soc. bras. Med.trop., v. 36, n. 3, p.349-353, 2003.

PASTORE, J.; HALLER, A. (1993) "What is happening to social mobility in Brazil?" In Albuquerque, R.; Velloso, J.P. R. (Orgs.). *Poverty and social mobility*. São Paulo: Nobel, p. 25-52.

POLIS, G. A. The Biology of Scorpions California: Stanford University Press, p.587, 1990.

PRENDINI, L.; WHEELER, W. C. Scorpion higher phylogeny and classification, taxonomic anarchy, and standards for peer review in online publishing. *Cladistics*, v. 21, n. 5, p.446-494, 2005.

SALLUM, A. M. C. et al. *O enfermeiro e as situações de emergência*. 2 Ed. Editora Atheneu: São Paulo, 2010.

SANTOS, M. Por uma outra Globalização: do pensamento único à consciência universal. 12 Ed. Rio de Janeiro: Record, 2005.

SELMA, T. S. [et al.].*Scorpions, spiders and snakes: general aspects and species of medical interest in the State of Alagoas*. Maceió: EDUFAL, 2005.

SILVA, A. F. Epidemiological aspects of accidents by venomous animals between 2007 and 2014 in the municipality of Palmas - TO. Porto Nacional (Bachelor's Degree in Biological Sciences) - Federal University of Tocantins -2015.

SOARES, M. R. M.; Azevedo, C. S.; Maria, D. M. Scorpionism in Belo Horizonte, MG: a retrospective study. Rev. Soc. Bras. de Med. Trop., v. 35,p. 359-363. 2002.

TORRES, J. B. et *al. Tityus serrulatus accidents and their epidemiological complications in Rio Grande do Sul. Rev. Saúde Pública*, v.36, n.º5, *São Paulo*, 2002.

VELLARD, V. Scorpions. Scientific mission to Goyaz and the Araguaia River. *Memoirs of the Zoological Society of France*, v. 29, n. 6, p. 539-556.

ZIEGLER, T. & W. R. LOURENÇO 2002. New scorpion records from the Gran Chaco of Paraguay (Chelicerata, Scorpiones). *Entomologische Mitteilun genaus dem Zoologischen Museum Hamburg*, 14(166): 63-69.

Websites

Palmas City Hall. Available at: <http://www.palmas.to.gov.br/>.

www.geocities.ws

http://www.skyscrapercity.com/showthread.php?t=1707463&page=2

yes

I want morebooks!

Buy your books fast and straightforward online - at one of world's fastest growing online book stores! Environmentally sound due to Print-on-Demand technologies.

Buy your books online at
www.morebooks.shop

Kaufen Sie Ihre Bücher schnell und unkompliziert online – auf einer der am schnellsten wachsenden Buchhandelsplattformen weltweit! Dank Print-On-Demand umwelt- und ressourcenschonend produzi ert.

Bücher schneller online kaufen
www.morebooks.shop

Printed by Books on Demand GmbH, Norderstedt / Germany